Mighty Machines

CRANES

Amanda Askew

QEB Publishing

Words in **bold** can be found
in the Glossary on page 23.

Copyright © QEB Publishing, Inc. 2010

Published in the United States by
QEB Publishing, Inc.
3 Wrigley, Suite A
Irvine, CA 92618

www.qed-publishing.co.uk

A CIP record for this book is available from the Library of Congress.

ISBN 978 1 59566 924 7

Printed in China

Written by Amanda Askew
Designed by Phil and Traci Morash (Fineline Studios)
Editor Angela Royston
Picture Researcher Maria Joannou

Associate Publisher Zeta Davies
Editorial Director Jane Walker

ISBN 978 1 59566 804 2 (paperback)

Printed in China

Picture credits
Key: t = top, b = bottom, c = center, FC = front cover, BC = back cover

Corbis Macduff Everton 13r, Reuters 17b; **Getty Images** AFP/Dimitris
Dimitriou/Stringer 14–15, AFP/Stringer 15c, 22bc; Photo courtesy of **The
GGR Group**/ GGR UNIC 21c; **Istockphoto** Dblight 9; **Rex Features** Sipa
Press 16–17, 22tc; **Shutterstock** Semjonow Juri FC, 1, 18–19, 22br, BC, Mihai
Simonia 4–5, 22bl, JoLin 5t, Ken Brown 6–7, Yui 8–9, Christian Lagerek 10b,
James Steidl 10–11, 22tr, Gary718 12–13, 22tl, Sergey Kozoderov 19r, Yantai
Raffles Shipyard Ltd 20–21, Rade Kovac 21c (background)

Contents

What is a crane?

Cranes are large machines that lift heavy **loads**. They can move things from side to side, and up and down. Cranes work mainly on **building sites**.

A crane can reach very high places. It can help to build skyscrapers.

There are many different types of crane. Some are used on water, some in the air, and some on land.

These two large cranes are working on a river.

Parts of a **crane**

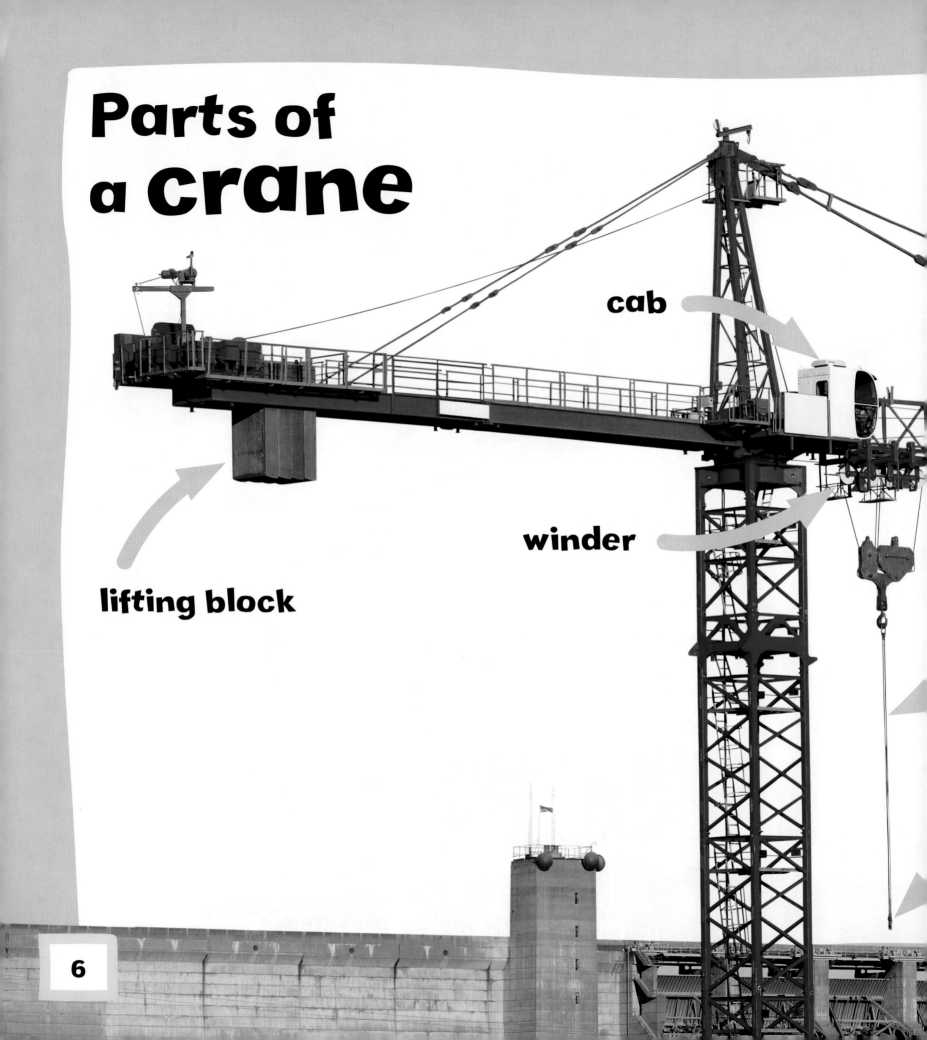

cab

winder

lifting block

6

A crane has three main parts—a cab, a **jib**, and a **pulley**. The cab is where the driver sits to control the crane. The driver has to climb up a long way to reach the cab.

pulley

jib

steel rope

The load is attached to the end of a steel rope. When the crane driver wants to lift the load, he winds in the rope.

hook

Lifting a weight

The lifting block balances the load and so stops the crane from falling over. The lifting block is also called the **counterweight**.

This lifting block moves along the shorter arm of the jib.

Some cranes have a sling instead of a hook. The sling can carry several things at the same time.

The lifting block moves in or out to balance the weight of the load. When the crane pulls the load in, the lifting block moves in, too.

Building up high

The higher the building, the taller the tower crane!

Tower cranes are the tallest cranes on building sites. A tower crane is fixed to the ground. It is put together on the building site.

The cab is right at the top of a tower crane. The driver climbs up several long ladders to reach the cab.

Tower cranes are so tall, someone has to tell the driver just where to put the load.

At the docks

A crane is used to load and unload ships. The crane loads **containers** and **crates** onto the ship. When a full ship arrives, the crane unloads its containers onto a truck.

The crane reaches right across the ship. It can lift heavier weights than other machines.

Most cranes that work in the **docks** move on rail tracks. The cranes lift containers from one fixed place to another.

The crane lifts a huge container onto the back of a lorry. Each lorry can carry just one container.

Building bridges

Some cranes are fixed to ships. They are called floating cranes. They can even work in rough, stormy seas. Floating cranes help to build bridges across rivers and lakes.

A floating crane lifts the last piece of the new bridge into place.

The biggest cranes can lift a piece of bridge that weighs about 880 tons (800 tonnes) —that's the same as 130 elephants!

This crane is lifting the wreck of a fishing boat from the seabed.

In the sky

Some helicopters work like cranes. They are called skycranes. They can move things to places other cranes can't get to. Skycranes often help during **disasters**, such as a **wildfire**.

The helicopter sucks in water through this pipe. The water is stored in a tank inside the helicopter.

When there is a wildfire, skycranes carry huge amounts of water over the fire. They drop the water onto the fire to put out the flames.

The water drops from the tank onto the fire.

Moving around

A mobile crane is fixed to the back of a truck. A mobile crane can be moved easily from one place to another. It is used to do quick jobs on a building site.

A mobile crane can travel along ordinary roads and streets.

A mobile crane has a **telescopic jib**. The jib can be made longer and longer to reach high up. When the crane has finished its work, the jib is made short again.

Biggest and smallest

The strongest crane in the world is in China. It is called Taisun and it helps to build huge ships. It can lift a load that is as heavy as 10,000 cars in one go!

Taisun is lifting part of an oil rig that is being built in the **shipyard**.

Mini cranes work in areas where there isn't much space. The smallest crane looks like a spider. It is called a crawler crane, and it is only 24 inches (60 centimeters) wide.

Although this crane is small, it can lift one ton —that is the same weight as 30 children.

Activities

- Here are three cranes from the book. Can you remember what they do?

- If you needed to lift a statue to the top of a very high building, which crane would you choose, and why?

- Draw a crane building a bridge. Which crane did you choose? What color is it? Who is driving it?

- Which picture shows a floating crane?

Glossary

Building site
A place where a house or other building is being built.

Container
A large box used to carry things on ships, trucks, and trains.

Counterweight
A heavy weight used to balance the load that the crane is lifting.

Crate
A large box made of wood or plastic.

Disaster
An event, such as a wildfire or a flood, that causes great damage.

Docks
A place where ships are loaded and unloaded.

Jib
The long arm of a crane.

Load
Whatever is being carried.

Pulley
Part of a crane where a rope moves over a wheel. A pulley makes it easier to lift heavy things.

Shipyard
A place where ships are built.

Telescopic jib
Made of separate parts that slide into each other to make the jib longer or shorter.

Wildfire
A fire that starts outdoors and spreads fast.

Index